International Conference on Communication Engineering and Logistics Management (ICCELM 2021)

IOP Conference Series: Materials Science and Engineering
Volume 1179

Online
23 July 2021

Editors:

Angle Liu
Iris Zhang

ISBN: 978-1-7138-4639-0
ISSN: 1757-8981

Printed from e-media with permission by:

Curran Associates, Inc.
57 Morehouse Lane
Red Hook, NY 12571

Some format issues inherent in the e-media version may also appear in this print version.

Institute of Physics
Dirac House, Temple Back
Bristol BS1 6BE UK

Phone: 44 1 17 929 7481
Fax: 44 1 17 920 0979

techtracking@iop.org

Additional copies of this publication are available from:

Curran Associates, Inc.
57 Morehouse Lane
Red Hook, NY 12571 USA
Phone: 845-758-0400
Fax: 845-758-2633
Email: curran@proceedings.com
Web: www.proceedings.com

TABLE OF CONTENTS

IOP Conf. Series: Materials Science and Engineering 1179 (2021) 011001 doi:10.1088/1757-899X/1179/1/011001

International Conference on Communication Engineering and Logistics Management[ICCELM2021], 23 July, 2021

1. About ICCELM2021

Yaseen Academy organized the **2021 International Conference on Communication Engineering and Logistics Management (ICCELM2021)**, which was sechduled to be held in during 24-26 July, 2021. Due to the restrictions caused by Covid-19, the conference was held in 23 July, 2021 online via Tencent Meeting. The Conference looks for significant contributions to all major fields of Communication Engineering and Logistics Management in theoretical and practical aspects. The aim of the conference is to provide a platform to the researchers and practitioners from both academia as well as industry to meet and share cutting-edge development in the field.

Organizer: *Yaseen Academy* is located at Huangjiahu Lake University Town, Wuhan, Hubei, China, focusing on international academic conferences organizing and international academic journals corperation in areas of science and engineering, aiming to provide a high-level and open academic platform for global scholars, researchers and practioners. It has gathered an academic committee composed of more than 200 experts and scholars, and has successfully held many international academic conferences. Facing the vast number of experts and scholars, Yaseen Academic is a team with strong professional quality and professional ethics, it has creating a mature, deep and complete academic exchange platform, not only allowing experts and scholars to show and share their latest academic research results and realizing face-to-face academic exchanges, but also allowing the most cutting-edge scientific information, ideas, and viewpoints to be fully communicated and exchanged, which will benefit all mankind.

2. Call for papers

The conference topics include, but are not limited to the following areas:

Information and Communication Engineering; Signal and Information Processing; Computer Science and Technology; Automation; Electronic Information Engineering; Software Engineering; Information Display and Opto-Electronic Technology; Intelligence Science and Technology; Electrical Engineering / Logitics Engineering, Engineering Management, Manufacturing Management, etc.

3. Conference Committee Members

Prof. Gangmin Li, Information and Computing Science, CSSE, Xi'an Jiaotong-Liverpool University(XJTLU), Suzhou, China / Member of IEEE, IEEE Computer society

Prof. Cai, Ning, School of Automation Beijing University of Posts and Telecommunications, Beijing, China

Prof. Dr. Chi-Hua Chen, College of Mathematics and Computer Science, Fuzhou University, China

Dr. Warayut Kampeerawat, Faculty of Engineering, Khon Kaen University, Khon Kaen, Thailand

Dr. Danish Khan, Electrical Engineering Department, Indus University Karachi, Pakistan

Prof. Alaa M. Rashad, Housing & Building National Research Center (HBRC), Cairo, Egypt

Prof. Tingrui Liu, College of Mechanical & Electronic Engineering, Shandong University of Science & Technology, Qingdao, China
Dr. Hassanien, Reda Hassanien Emam, Cairo University, Cairo, Egypt
Dr. Saad Ahmed Javed， Nanjing University of Information Science and Technology, Nanjing, China
Dr. Yanbin Wang, Nanjing University of Aeronautics and Astronautics, Nanjing, China
Dr. Vijay kumar, Faculty of Engineering & Technology in Manav Rachna International Institute of Research & Studies (Formally Manav Rachna International University), Faridabad, India
Dr. Elisa Marrasso, University of Sannio, Benevento, Italy
Dr. Maryam EL MAROUANI, Department of chemistry, Hafr Al Batin University, Saudi Arabia
Dr. Raghabendra Yadav, Thapathali Campus, Tribhuvan University, Nepal
Dr. Fanlong Zhang, School of Computer, GuangDong University of Technology, China
Dr. Mohammad Esmaeili, Department of Electrical Engineering, University of Texas, Dallas, Texas, USA
Dr. Syed Abdul Rehman Khan, Department of supply chain and logistics, Xuzhou University of Technology, China

Editors:
Angle Liu, Hankou University, Jiangxia District, Wuhan, Hubei, China, anqiliu0615@163.com
Iris Zhang, Yaseen Academy, Hongshan District, Wuhan, Hubei, China, iriszhang@yaseen.org.cn

4. Paper Submission
The total count of paper submission was 10. A preliminary review was conducted by the main editor and only the approved papers were sent for peer-reviewed. The peer review for selection of papers was conducted then, each paper was reviewed by reviewed by 2-4 reviewers(average three) and finally 3 papers were accepted for paper presentation and publication in the conference. The acceptance rate is 30%.

5. Programs
The conference was bagan at 14:00, 23 July, 2021 and Iris Zhang of Yaseen Academy hosted the conference. There were 9 participants in total. One invited keynote speaker delivered his speech, the speech cost about 50 minutes and 5 minutes for questions&discussion. And the other two authors share their papers after the keynote speech, each sharing cost about 10 minutes.

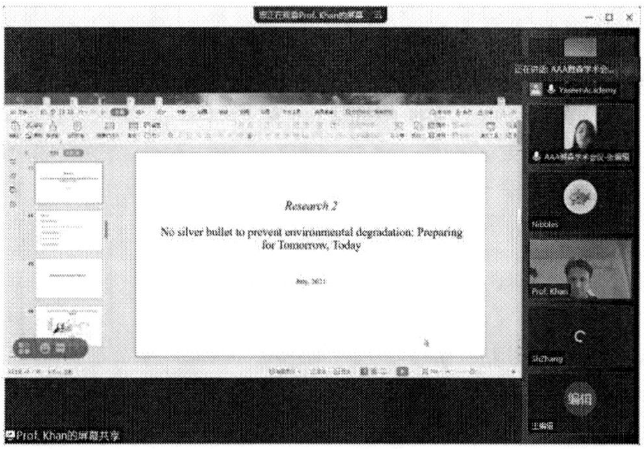

Figure 1: Prof. Dr. Syed Abdul Rehman Khan is delivering the speech
Speech Title： No silver bullet to prevent environmental degradation:Preparing for Tomorrow,Today

Prof. Dr. Syed Abdul Rehman Khan
Department of supply chain and logistics, Xuzhou University of Technology, China

Biography: Dr Syed Abdul Rehman Khan is an expert of Supply Chain and Logistics Management. Dr Khan achieved his CSCP—Certified Supply Chain Professional certificate from the U.S.A. and completed his postdoctoral fellowship from Tsinghua University. Since 2018, Dr Khan is a Professor of Logistics and Supply Chain Management. He has more than twelve years' core experience of supply chain and logistics at industry and academic levels. He has attended several international conferences and also has been invited as a keynote speaker in different countries. He has published more than 120+ scientific research papers in different well-renowned international peer-reviewed journals (SSCI/SCI and ABS listed) and conferences, including couple of research paper indexed in Essential Science Indicators (ESI). Dr Khan is the authored of 7 books related to the sustainability in supply chain and business operations. He is a regular contributor to conferences and workshops around the world.

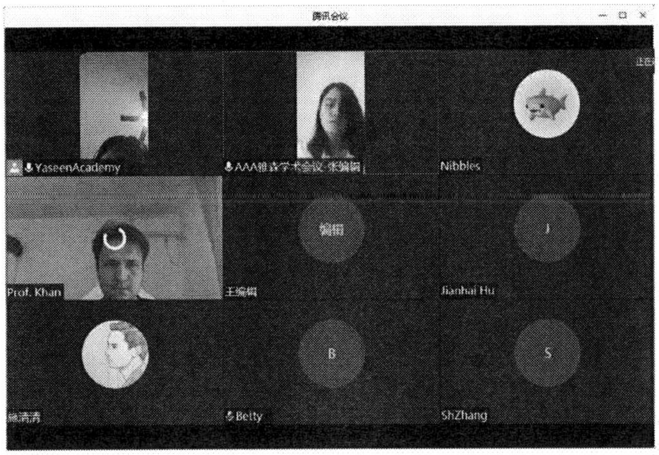

Figure 2: The screen Shot during the conference

Peer review declaration

All papers published in this volume of IOP Conference Series: Materials Science and Engineering have been peer reviewed through processes administered by the Editors. Reviews were conducted by expert referees to the professional and scientific standards expected of a proceedings journal published by IOP Publishing.

- **Type of peer review:** Double-blind review
- **Conference submission management system:** Yaseen Academy
- **Number of submissions received:** 12
- **Number of submissions sent for review:** 10
- **Number of submissions accepted:** 3
- **Acceptance Rate (Number of Submissions Accepted / Number of Submissions Received X 100):** 30%
- **Average number of reviews per paper:** 2
- **Total number of reviewers involved:** 5
- **Any additional info on review process:** A preliminary review was conducted by the main editor and only the approved papers were sent for peer-review.
- **Contact person for queries:** Iris Zhang (Yaseen Academy, Hongshan District, Wuhan, Hubei, China, iriszhang@yaseen.org.cn)

Baseline Test Suite Construction of Smoke Test for Extreme Programming

Mengqing Tanli[1,*], Ying Zhang[2], Yan Jiang[2] and Yulin Wang[2]

[1]School of Software, University of South China, Hengyang, Hunan, China
[2]School of Mechanical Engineering, University of South China, Hengyang, Hunan, China

[*]corresponding author's e-mail: TLMQ-TEAM@163.com

Abstract. This paper proposed an approach of construction of baseline test suite in smoke test for extreme programming. According to our investigation, the baseline test suite should firstly embody the thinking of "data-centerlization", and the baseline test suite in smoke test should be conducted in terms of the true application scheme, so this baseline test suite should be the calibration marking for testing and verification of a baseline product version. As a typical example, in the procedure of construction in smoke test of Product Quality Monitoring Software, data accuracy, kinds of control chart, data range, typical application situation, and further requirement were taken into consideration. In order to improve the testing efficiency and to avoid mistakes, a loading kit was developed.

1.Introduction

Extreme programming is excellent programming strategy for varied requirement. We know, requirement is changing with the user, but the objective of the software product is relatively static in a survival period and the application scheme of the software product will keep stationary. So, we can construct test suite according to this true application scheme of the software product especially for extreme programming, in which, all kinds of test data will embrace these true application schemes, as shown in figure 1. Consequently, this kind of test suite will become a calibration marking in smoke test for testing the baseline product version of the software [1-3].

Additionally, in engineering and business activity, although business process will be distinguished for all kinds of factual object, basic features of business data will always keep relative invariant for the same subject. Not only taking the actual changed requirement into consideration but also giving and using an advantage of our experience, data-centerlization would be a good mode for software testing in extreme programming, especially in smoke test [4, 5].

1.1. Investigation of baseline test suite in smoke test

Extreme programming is kind of test-driven programming that test case and test suite is firstly designed in programming for the changed and specific requirement, and testing is taken as a core fact for software programming. For an update software product version, the smoke test is the primary step in testing activity [6]. So, the construction of baseline test suite in smoke test is very important for consequent software testing and programming. Of course, as we known, smoke test must firstly determine main function of the software product. As a result, following discussion is beginning with the analysis of software main function using IDEF0 diagram.

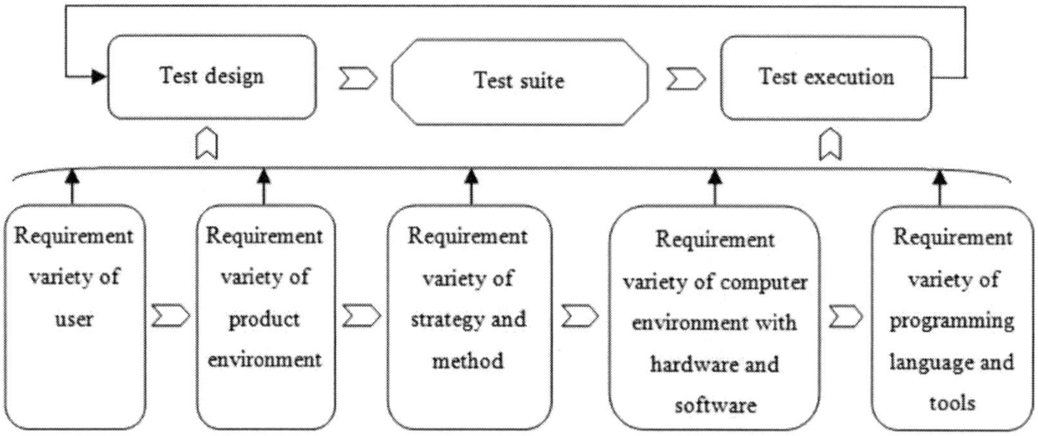

Figure 1. Software testing for extreme programming

1.1.1. Analysis of software main function in smoke test for baseline version. The IDEF0 diagram for main function analysis of Product Quality Monitoring System Version 2.0 (abbreviation is PQMS2) is shown in figure 2 [7, 8]. And according to figure 2, we can know that main function of the software is inputting of inspection data, computing and drawing control chart, and we can give the basic disposing mode according to domain expert of quality control as following.

- Mode1 - Manually inputting inspection data, computing and drawing control chart.
- Mode2 - Inputting inspection data, computing and drawing control chart by digital gauge and measure tools.
- Mode3 - Inputting inspection data, computing and drawing control chart using saving data.
- Mode4 - Inputting inspection data, computing and drawing control chart using Notepad of Windows.

1.2. Affecting factors of baseline test suite based on true application scheme [9-14]

For true application scheme, a lot of test data must be generated in baseline test suite. These test data should not be disordered and disorganized [9, 10]. Therefore, the construction of baseline test suite should considerate various influence factors [11-14], and factors for true application scheme mainly include:

1.2.1. Data accuracy. For counting value data, data value is integer, and data accuracy is 1.

a. For counting value data, data value is integer, and data accuracy is 1.
b. For measuring value data, data accuracy usually access 0.1, 0.01, and 0.001 in manufacturing factory.

1.2.2. Kinds of control chart. Generally, control chart include two types — the control chart of measuring value data and the control chart of counting value data. Among control chart of measuring value data, there are three kinds — Median control chart, Average value control chart and Individuals control chart. And among the control chart of counting value data, there are four kinds — NP control chart, P control chart, C control chart, and U control chart.

a. Median control chart/ Median and range control chart

As a control chart of measuring value, the Median control chart is fitted to inspection data of little range value. And it is popularly used for mass and large batch production.

b. Average value control chart/ Average and moving range control chart

The average value control chart is a typical control chart of measuring value, and it is widely applied in quality control domain for mass and large batch production.

c. Individuals control chart/ Charlie control chart
Individuals control chart is generally used in the state which the inspection time is very long or the inspection data is difficultly gathered, for example, the specimen / work piece would be destroyed in test processing.

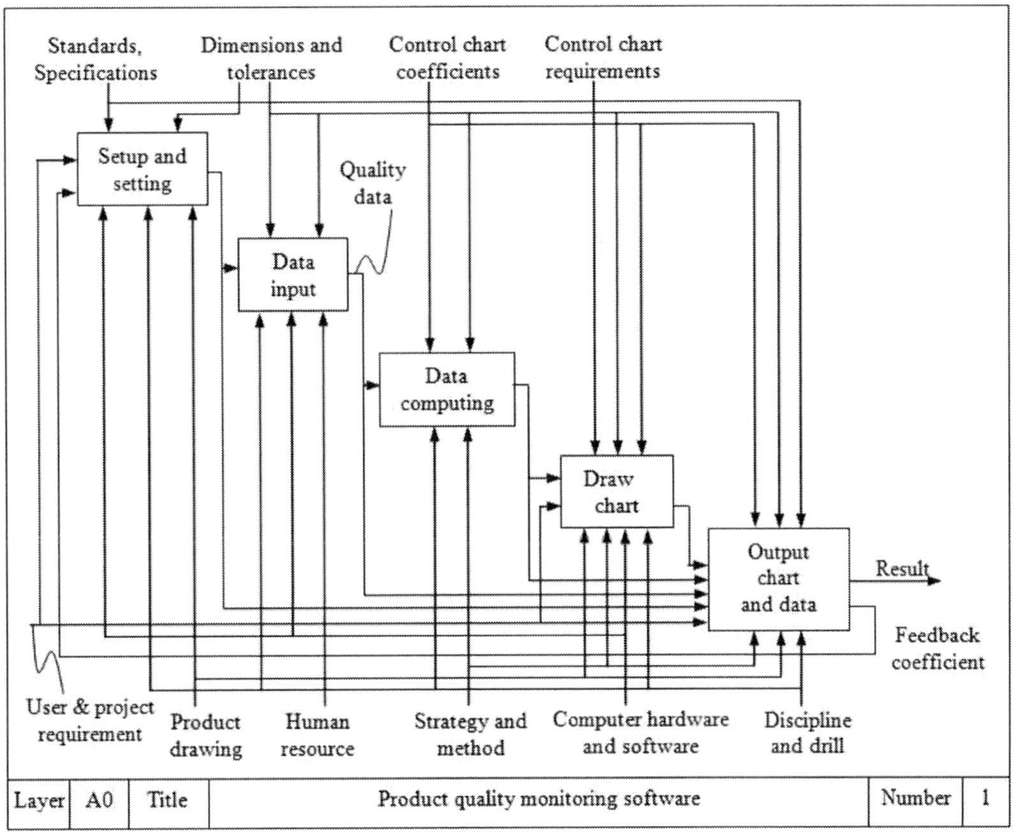

Figure 2. IDEF0 of software main function

d. NP control chart
NP control chart is a kind of control chart to monitor numbers of unqualified product. This control chart is fitted to acceptance testing.

e. P control chart
P control chart is a kind of control chart to monitor rates of unqualified product. . This control chart is also fitted to acceptance testing.

f. C control chart
C control chart is a kind of control chart to monitor numbers of defect/fault, and it is used for testing of point counting value, especially for numbers of flaw.

g. U control chart
U control chart is a kind of control chart to monitor rates of defect/fault or defect/fault per unit, and it is also used for testing of point counting value, especially for the rate of flaws.

1.2.3. Data range. For data type of computer programming, there are its own definition and constraint according to programming language and hardware configuration. But for typical application scheme, we should pay attention to following specifications: (1) for measuring value data, general true value is among 0 and 1000, and most is not beyond 500; (2) for measuring value data, general deviation value is among 0.000 and 10.000, and most is less than 1.000; (3) for counting value data, general number of specimen is among 0 and 10000, and most is not beyond 1000.

1.2.4. Typical application

a. Purchase inspection

Purchase inspection executes the testing for purchasing, including all testing of material, standard part, and special part or component, etc.

b. Process test

Process test includes mainly two aspects, that is, process testing of fabrication part for factory self, and testing of assembly process.

c. Finish product test

Finish product test is the end test of a product in a factory.

1.2.5. Further requirement. For extreme programming, the requirement of user for software product will be multi-layer and multi-dimension. Sometimes, there are some further special requirements for test-driven testing. For example, in PQMS2, distinguished amplification coefficient is required for R chart different from XAve chart in the chart when the range value is very tiny.

2. Construction of baseline test suite of smoke test for extreme programming

2.1. Criterion and procedure of construction of baseline test suite for smoke test

At first, in smoke test for baseline version, construction and arrangement of the test suite must be based on of main function to fulfill the aim of calibration marking. And then, details should be conducted in terms of affecting factors of application scheme to assure that baseline test suite is effective, completed, systematical and convenient to use.

Consequently, construction and arrangement of the test suite must be obliged to the standard and specification of software producing and testing. Disposing procedure of test suite in a survival period should be deferred to figure 3.

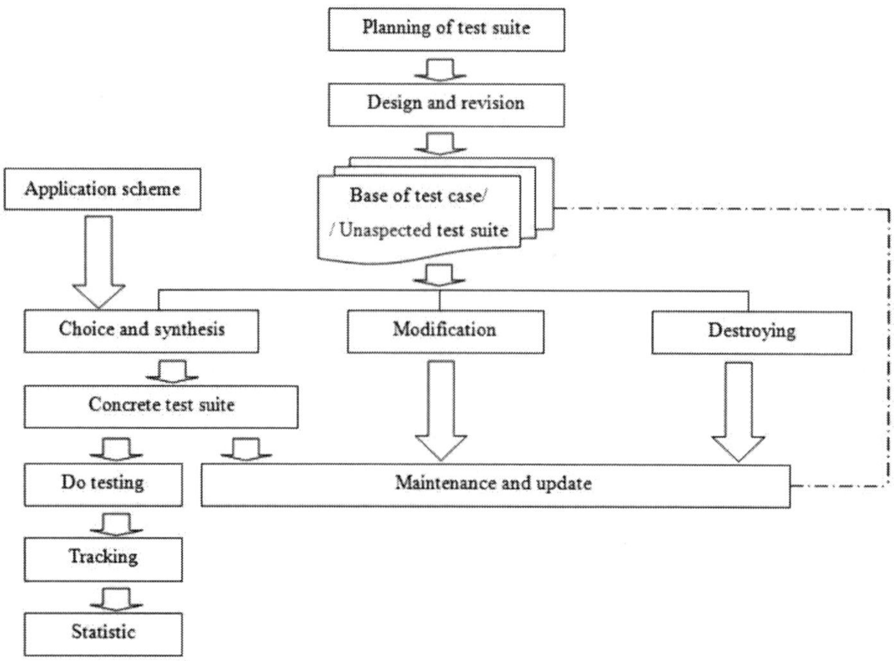

Figure 3. Disposing procedure of test suite in a survival period

2.2. Concrete contents of baseline test suite of PQMS2

2.2.1. Arrangement of typical application scheme. Typical application scheme mainly focus on quality control phrases with respect to all kinds of control chart [15]. The table1 has demonstrated the arrangement about this issue. In table 1, columns include Median (control chart of median and range value), XAve (control chart of average and range value), X (control chart of individuals and moving range), NP (control chart of numbers of unqualified product), P (control chart of rate of unqualified product), C (control chart of numbers of defect), and U (control chart of rate of defect), while rows include three typical application schemes.

Table 1. Arrangement of typical application scheme in baseline test suite.

	Median	XAve	X	NP	P	C	U	Sum	Percent
Purchase inspection	1	2	-	-	-	-	-	3	27.2%
Process test	3	-	-	-	1	1	1	6	54.6%
Finish product test	-	-	1	1	-	-	-	2	18.2%

As the table 1 shows, the rate concerning process test achieves 54.6%, the rate of purchase inspection gains 27.2%, and the rate about finish product test is 18.2%. We can find that the arrangement of all kinds control chart is reasonable for various quality control phrases.

2.2.2. Composition of test data in baseline test suite for smoke test [16-18]. Test data plays a very important role in software testing, and it is generally taken as a part of test case in baseline test suite. Particularly, the quality of test data will play very important influence in smoke test while the composition and state of test data will directly determine the effectiveness of true application scheme or situation. Table 2 has illustrated the composition and data volume of baseline test suite for smoke test.

Table 2. Data volume of baseline test suite for smoke test.

			Manually inputting	Inputting by digital gauge and measure tools	Inputting using saving data	Inputting using Notepad of Windows
Kinds of control chart	Measuring value	Median	60	70	60	40
		XAve	-	40	90	-
		X	-	12	-	-
	Counting value	NP	20	-	-	-
		P	-	-	-	16
		C	-	-	-	12
		U	-	-	-	12
Sum			80	122	150	60

As above table exhibits, for the former three modes of inputting data, data volume is relatively larger, because the first mode and the third mode can supply better safety of data with data verification of limitation and format, and the second mode would improve the efficiency of data acquisition. But the data volume of the fourth mode is relatively decreased because of safety of inputting data and lower efficiency. All these disposing will improve the effectiveness of baseline test suite far from virtual application scheme.

2.2.3. Constructing baseline test suite for smoke test considering affecting factors. According to analysis of the above section 2.2, we can do concrete and detail construction of baseline test suite for smoke test. The table 3 has demonstrated the concrete state in terms of data accuracy, kinds of control chart, and data range. In table 3, the symbol "★" imply that the factor at intersection position is turned into fact for conducting test case in baseline test suite for true application scheme. Still, the last column lists the special requirement -add marking for tiny range value.

Table 3. Construction of baseline test suite for smoke test in PQMS2.

				According to main functions				
				Manually inputting	Inputting by digital gauge and measure tools	Inputting using saving data	Inputting using Notepad of Windows	Further requirement -add marking for tiny range value
According to affecting factors	Data accuracy		0.001	-	★	-	-	-
			0.01	-	★	-	-	★
			0.1	-	★	-	-	-
	Kinds of control chart	Measuring	Median	★	★	★	★	-
			XAve	-	★	★	-	★
			X	-	★	-	-	-
		Counting	NP	★	-	-	-	-
			P	-	-	-	★	-
			C	-	-	-	★	-
			U	-	-	-	★	-
	Data range		10-2 range	-	-	★	-	-
			10-1 range	-	-	★	-	★

In table 3, it implies that all factors must have its representative and should indicate the factual application scheme with fewer unmeaning repeat as possible.

2.3. Loading kit of basic test data for smoke test in PQMS2

The baseline test suite usually includes a lot of test data, and there are many basic test data for setting up of testing environment in smoke test, while some test data are remaindering or unmeaning. If these basic test data all are treatment manually, it will decrease the efficiency of testing. So, a loading kit is facilitated for initialization of testing environment. Basic test data organization of loading kit in smoke test is shown in figure 4. It mainly includes three sections that factory divisions embodies the general organization of factory, product and part mainly use factual and meaningful subjects, and inspection processes focus on key control point of product quality. Of course, these basic data also imply the true application scheme [15].

For programming of this loading kit, this technical task should performed by testing engineer because of the convenience of using and maintenance. In order to fulfill this situation, the testing engineer must be disciplined and drilled perfectly. On one hand, testing engineer must take part in the developing procedure and keep cooperation with the programmers. On the other hand, it is required that testing engineer should be familiar to the software programming and be knowledgeable of usual computer programming language [8].

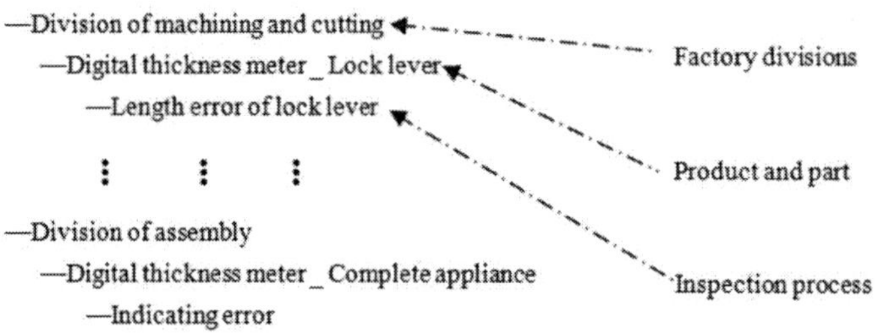

Figure 4. Basic test data organization of loading kit in smoke test

Against the maintenance, because the test data will be changed with the update of true application scheme and baseline product version, the testing engineer must edit test data following this change.

3. Result and conclusion

For extreme programming, the software programming is driven by testing. As a result, the construction of baseline test suite becomes a very important working, in which, the test case must be updated according to the variety of all kinds of requirement. But, for a baseline version of a software product, the true application scheme of the software generally keeps stationary, so the baseline test suite in smoke test should be conducted in terms of this application scheme, and this baseline test suite will be the calibration marking for testing and verification of a baseline product version.

Based on the true application scheme, we have constructed a baseline test suite of smoke test for PQMS2, while following five aspects have been taken into consideration, that is, data accuracy, kinds of control chart, data range, typical application, and special requirement. Consequently, a test loading kit is facilitated for PQMS2. With the installation of loading kit of basic test data, the efficiency of testing will be promoted and the unnecessary mistakes will be avoided in smoke test.

References

[1] Jeremy S. Bradbury, James R. Cordy, Juergen Dingel 2005 Proc. Int. Conf. of the ACM SIGPLAN-SIGSOFT Work on Program Analysis for Software Tools and Engineering vol 1. (Portugal: Lisbon) pp 160-170.
[2] Per Runeson 2009 J. Empirical Software Engineering. 14 2 131–164.
[3] B. W. Boehm 1979 Classics in software engineering (New Jersey: Yourdon Press Upper Saddle River).
[4] Kent Beck 2003 Test-Driven Development by Example (China Electric Press).
[5] S McConnell 1996 J. IEEE Software. 13 4 143-144.
[6] B. W. Boehm 1981 Software Engineering Economics (Upper: Prentice-Hall).
[7] Information on http://www.baidu.com, 2020-2021.
[8] MENGQING TANLI, YAN JIANG, YULIN WANG 2020 Proc. Int. Workshop on Computer Science and Engineering vol 1(Shanghai, China) pp 18-26.
[9] XU Yuan-yuan 2015 J. Software. 36 9 117-120.
[10] Tang Lingyao 2005 D. of National University of Defense Technology (Changsha, Institute of Computer Science) .
[11] Fu Bing 2014 Course of Software Testing (Tsinghua University Press).
[12] Ma Xueying 2008 D. of ZheJiang University (Hangzhou, College of Computer Science).
[13] Ron Patton 2006 Software engineering (USA: Pearson Education Inc.).
[14] Capers Jones 1997 Software Quality: Analysis and Guideline for Success (International Thomson Press).
[15] Dan Tang, Mengqing TanLi, Yan Jiang 2019 Proc. the 4th Annual Int. Conf. on Information System and Artificial Intelligence, vol 1 (Changsha) p123-130.
[16] Mengqing TanLi, Yan Jiang, Xiang Wang 2019 Proc. Int. Conf. on Computer Science, Communication and Network Security vol 1 (DEStech Publications, Inc.) p 198-204.
[17] TANLI Meng-qing, ZHANG Ying, WANG Yu-lin 2020 J. Software. 41 9 5-8, 25.
[18] TANLI Meng-qing, ZHANG Ying, WANG Yu-lin 2020 J. Software. 41 11 1-4, 25.

IOP Conf. Series: Materials Science and Engineering 1179 (2021) 012002 doi:10.1088/1757-899X/1179/1/012002

Design of Intelligent Early Warning Robot System for Warehouse Autonomous Patrol based on Scale Estimation Algorithm of Probability Theory

Jie Li[1], Yaoyao Tu[1,*] and Biao Lu[2]

[1]Faculty of Mathematics and Statistics, Suzhou University, Suzhou, Anhui, China
[2]Faculty of Information and engineering, Suzhou University, Suzhou, Anhui, China

[*]corresponding author's e-mail: tyy20101112@163.com

Abstract. In view of the backward monitoring status in the traditional safety management of hazardous chemicals warehouse, a patrol robot for hazardous chemicals warehouse with autonomous patrol stereo detection, intelligent early warning and remote monitoring functions is designed, and the design process of related functions of autonomous patrol navigation subsystem, environmental detection subsystem and remote monitoring subsystem is discussed from two aspects of hardware and software. Because both the camera and the target may be in motion, it is necessary to consider the scale change of target influence caused by relative distance. In order to overcome the shortcoming of invariable scale in Mean Shift algorithm, this paper adopts a scale adaptive method to estimate the size of the target, adjusts the parameters of kernel function by using ellipse feature, and takes the parameters of the current frame as the initial data of the next frame, so as to achieve the effect of adaptively adjusting the target scale. The test results show that the system can effectively traverse the map of hazardous chemicals warehouse, accurately assess the leakage of hazardous chemicals and the environmental safety of warehouse, and give early warning. The remote monitoring platform greatly facilitates the remote management of the warehouse.

1. Introduction

With the development of science and technology and the progress of society, the warehousing industry has entered a high-speed development stage, and the warehouse space area is getting larger and larger, but the warehouse supervision technology has not been developed synchronously, which leads to frequent dangerous accidents in warehouses everywhere. Chemical and chemical industry is an important part of China's economy, and hazardous chemicals account for a considerable share of chemical products[1]. Dangerous accidents in hazardous chemicals warehouses are mostly caused by changes in warehouse environment, which is coupled by variables such as air temperature and humidity, pressure, light, and concentration of leaked hazardous chemicals[2]. Detection of relevant environmental variables in the whole range of warehouses is a prerequisite for preventing dangerous accidents.

At present, there are two ways to manage hazardous chemicals warehouse: pure manual monitoring management and equipment monitoring management. However, both pure manual monitoring and equipment monitoring have great defects[3]. Under the pure manual management mode, it is very difficult to patrol without errors 24 hours a day, regardless of whether there are enough warehouse managers[4]. Moreover, because the human body is not sensitive enough to environmental information, it is not easy to detect when the warehouse environment deteriorates[5]. With the help of equipment monitoring, there are also the following problems: (1) monitoring blind areas are easy to exist in the fixed-point installation of monitoring devices in warehouses; The installed detection sensor is single, and the pertinence is not strong, the rate of missing alarm and false alarm is high, and the system lacks intelligence; Monitoring devices need to be installed at multiple points, and the system cost is too high[6]. In view of the above problems, this paper designs a patrol robot system with autonomous patrol, stereo detection, intelligent early warning and remote monitoring functions, which effectively makes up for the shortcomings in the safety monitoring of hazardous chemicals warehouses.

2. Overall scheme design

The patrol robot can patrol the hazardous chemicals warehouse autonomously in the whole range with the help of the patrol navigation subsystem. At the same time, it can detect and fuse various environmental parameters and leaked hazardous chemicals concentration in the warehouse through the environmental monitoring subsystem, and upload various original data and fused data to the cloud server for storage through WiFi wireless network[7]. At the same time, the robot will perform data fusion processing on all kinds of environmental data collected, and compare the data fusion result value with the alarm threshold value to judge the dangerous degree of the warehouse environment. When the robot thinks that there is a potential safety hazard in the warehouse comprehensive environment, it will send an early warning or even an alarm signal. When the manager fails to handle it in time, the robot will call the warehouse manager through GSM, and the staff can also log in to the cloud server through the remote monitoring subsystem to check the warehouse situation anytime and anywhere or send various control instructions to the robot[8]. The whole system structure is shown in Figure 1:

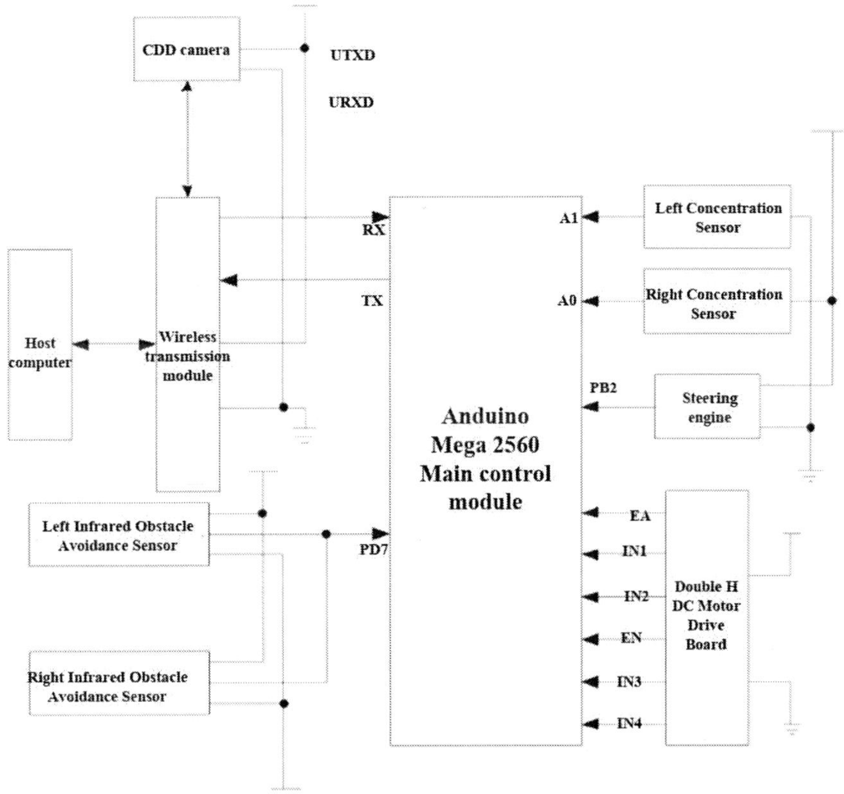

Figure 1. System structure diagram of patrol robot

3. System hardware design

3.1. Patrol navigation subsystem.

This subsystem mainly realizes the autonomous patrol function of patrol robot in indoor closed environment. In hardware, it consists of microcontroller STM32F429IG, minimum system circuit motor and drive module L298 N 6-axis motion processing module MPU 6050 infrared obstacle avoidance detection sensor E18-D80NK, etc. Firstly, L298N drive motor is used to make the robot move, and E18-D80NK sensor is used to detect obstacles. The built-in angular velocity sensor of MPU6050 is used to detect the direction angle of the robot during patrol[9]. After the robot patrols once, the patrol mode automatically switches between the right-hand rule and the left-hand rule. At the same time, PID algorithm is used to control the robot's traveling path during patrol, thus realizing the robot's autonomous patrol navigation function in the hazardous chemicals warehouse. The hardware structure diagram of patrol navigation subsystem is shown in Figure 2:

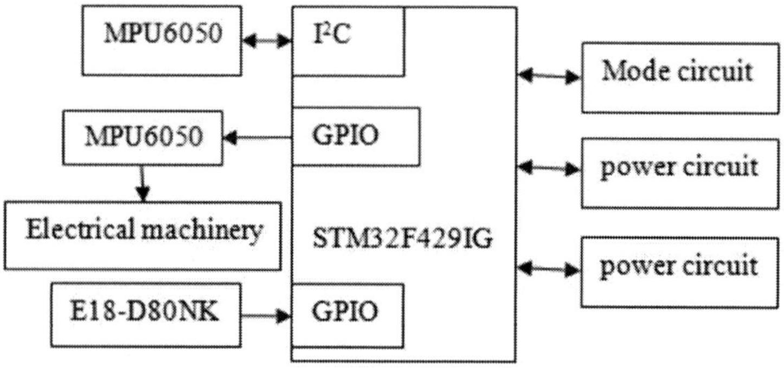

Figure 2. Hardware structure diagram of patrol navigation subsystem

3.2. Environmental detection subsystem

This subsystem mainly realizes the functions of multi-sensor data fusion, data communication and video monitoring for detecting various environmental variables around the robot, Hardware includes microcontroller STM32F429IG (shared with patrol navigation subsystem), gas detection sensor MQ2 (which can detect liquefied gas, alkane, benzene, hydrogen, alcohol smoke, etc.), flame sensor Rb-02s022a, temperature and humidity sensor DHT11, and other environmental monitoring sensors (if necessary, the required sensors can be added to the reserved MCU interface according to different needs)[10]. Communication module includes SIM808 and WiFi module RTL8188EUS, which integrate GSM bluetooth technology. video monitoring adopts the combination of electrically controllable pan/tilt and camera, and users can view images from all angles in the warehouse by remotely controlling the pan/tilt through the camera. On the working principle, it firstly detects the surrounding environment by using the above-mentioned sensors carried by the robot itself during the patrol process, at the same time, it uses MCU to fuse the data collected by each sensor with the multi-sensor data fusion model based on BP neural network, and judges the danger degree of the current warehouse environment by itself according to the fusion results[11]. When the fusion results approach the alarm threshold, the robot will automatically send corresponding early warning or even alarm signals to the remote monitoring platform through WiFi, and when the managers fail to process them in time, the robot will also use SIM808 to call the managers. The hardware structure diagram of the whole environment detection subsystem is shown in Figure 3:

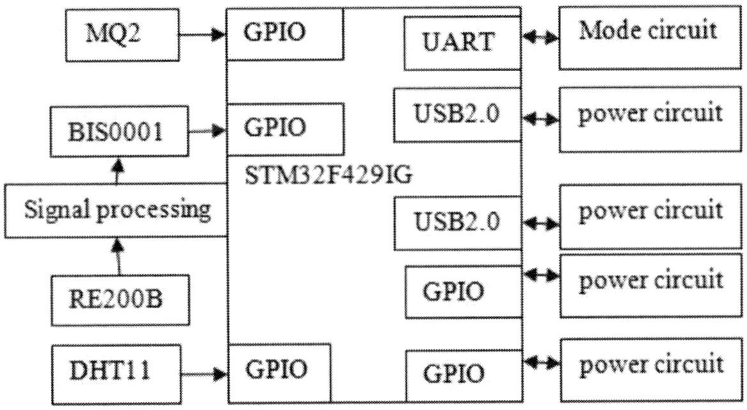

Figure 3. Hardware structure of environmental detection subsystem

4. System software design

4.1. Autonomous patrol function

Autonomous movement in indoor environment is the premise for robots to monitor the environment. In order to realize autonomous patrol of robots in indoor environment, infrared sensors are used to detect the boundaries of walkable areas, and the left-hand rule and the right-hand rule are used as the guidelines for robots to patrol the whole map. At the same time, the azimuth and mileage accumulated during the robot's forward movement are detected by angular velocity sensors and the number of turns of motors. After each turn, the robot switches patrol between the left-hand rule and the right-hand rule at 45 degrees, and the robot moves forward in parallel along the edge of indoor map with PID control algorithm. In this paper, the left-hand rule is defined: in the process of autonomous patrol, when an obstacle is detected on the left side (that is, the boundary of the walking area), the robot uses PID algorithm to keep a certain distance from the boundary and move along the boundary until it completes one turn. In the process of the robot moving forward, the distance traveled and the converted direction angle are continuously accumulated. When the distance is equal to the preset perimeter of the indoor open space patrol path and the steering angle of the robot orientation reaches 360, it is considered that the robot has completed one turn. The right-hand rule is similar. The flow chart of autonomous patrol algorithm used in this system is shown in Figure 4:

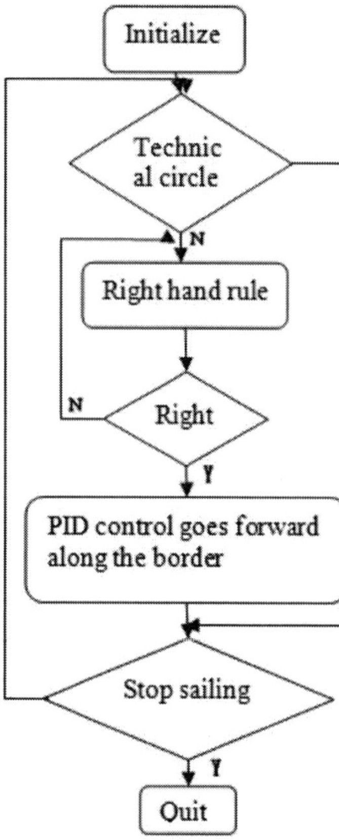

Figure 4. Flow chart of robot autonomous patrol algorithm

4.2. Multisensor data fusion function

It is a very important function of patrol robot to estimate the potential danger degree of the environment by detecting various parameters of the surrounding environment. It is the most commonly used detection method to sense the temperature, humidity, toxic, flammable and explosive gas concentration in the environment by carrying multiple special sensors. However, due to the coupling effect of various environmental variables in different degrees, for example, when the relative humidity in the air is greater than 60%, there are few fires, and the safe concentration threshold of flammable and explosive gases in the air is greatly improved; When the humidity is between 40% and 60%, it can smoke and burn, but it is not easy to expand the burning area. When the air humidity is between 30% and 40%, it is easy to burn and spread, while when the humidity is lower than 25%, it is easy to have a fire, and a very small amount of flammable and explosive gases in the air may also cause disasters. For this reason, the system makes use of the complementarity of multiple heterogeneous sensors to enrich the detection of various variables in the space environment, and designs a feature-level data fusion model with BP neural network as the core algorithm to evaluate the risk of environmental parameters.

The multi-sensor data fusion system based on BP neural network is designed in this paper. Its principle is to collect warehouse environment variables through N special-shaped sensors of patrol robot system, then denoise the collected data by using Laida rule, then normalize the data, and take the processing results as the standard input data of BP neural network. Finally, the trained neural network is used for fusion processing, and the fusion results are output as the basis of robot alarm decision.

The neural network model designed in this paper includes an input layer, a hidden layer and an output layer. For the convenience of research, this paper selects three typical environmental variables (including air temperature x1, air relative humidity x2, the comprehensive value of various toxic or flammable hazardous chemicals concentration x3) as the network input, and selects the early warning

IOP Conf. Series: Materials Science and Engineering 1179 (2021) 012002 doi:10.1088/1757-899X/1179/1/012002

variable y ($0 \leq y \leq 1$) as the output of the network output node.

4.3. Remote monitoring subsystem

On the one hand, during patrol, the robot periodically collects the surrounding environment data through the environment detection subsystem, fuses the collected data by using the BP neural network data fusion model, and then uploads the collected environment data (including the fusion results) to the cloud server through the gateway. When the fusion results are close to the alarm threshold, the robot will also send an early warning signal to the monitoring platform, and when the manager fails to handle it in time, the robot will also send an alarm signal to the administrator's mobile phone through GSM. On the other hand, managers can also access the server through the PC monitoring platform or mobile phone monitoring platform, view the current warehouse environment anytime and anywhere, and send instructions to the robot remotely through the monitoring platform if necessary.

The remote monitoring subsystem includes robot program, cloud server program, PC monitoring platform program and mobile phone App program. Among them, the robot program mainly realizes the functions of collecting environmental information, fusing data, uploading and sending alarm signals and receiving control signals. The cloud server is based on the PaaS Internet of Things open platform oneNET provided by China Mobile, and is mainly used to receive and save the data uploaded by the robot and provide data access services for PC monitoring platform and mobile App; The PC monitoring platform is implemented by using oneNET's supporting WEB service, which is mainly used to read the environmental information uploaded by the robot and the state parameters of the robot itself from the cloud server. In addition to the PC monitoring platform program function, the mobile App also adds the robot remote control function, and its workflow is shown in Figure 5:

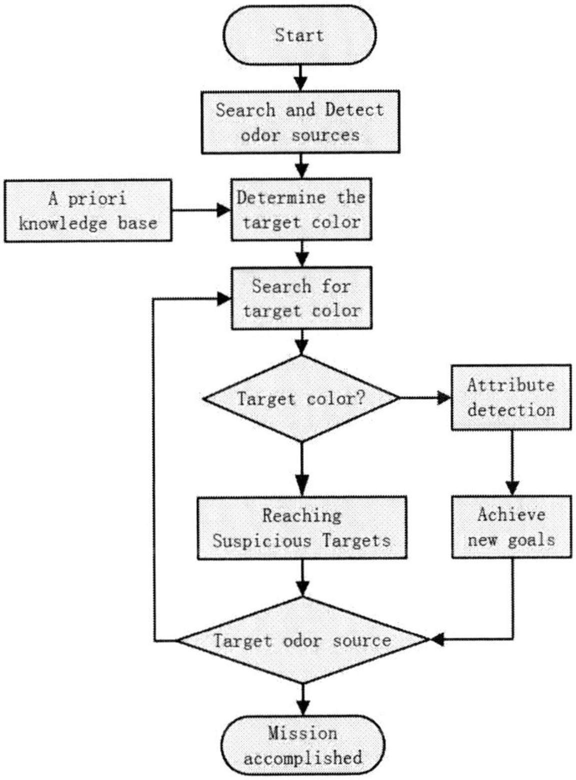

Figure 5. Work flow chart of remote monitoring subsystem

5. Mean Shift target tracking based on scale estimation

The size of the target changes with the movement. When the target object moves in the shooting range, because the search window is fixed and the size of the target itself becomes larger, some points of the target pixel will be excluded from the search window, resulting in inaccurate convergence of local maxima of sample points and tracking errors. When the target size becomes smaller, non-target pixels will be included in the search window, that is, non-sample particles will be locked, which will often lead to tracking failure. Therefore, this paper uses scale estimation to improve the tracking drift phenomenon caused by the constant window of traditional Mean Shift. In the tracking process, the tracking window is adaptively adjusted according to the change of target size.

5.1. Parameter adjustment of kernel function

In this paper, target color features are used as target input information. Based on the traditional Mean Shift, the ellipse which is the closest to the target description is established as the search area, which converges in the area. The template parameters are constantly updated by using the iterative idea of Mean Shift, and the target parameters of the previous frame are used as the initial parameters of the next frame, so as to achieve the purpose of adaptively adjusting the scale according to the change of the target. When using Mean Shift algorithm to track the target, firstly, the initialization setting is carried out, that is, the parameters of the system are set. The initialization methods can be divided into two types: automatic mode and semi-automatic mode. The biggest difference between the two methods lies in the different methods of obtaining the target area to be tracked. The automatic method is to automatically obtain the target area to be tracked by the method of target detection, while the semi-automatic method is to specify the interested target area in the original picture by manual method. In this study, a semi-automatic method is adopted, that is, the tracking target is selected manually[12].

Suppose $\{x_i\}(i=1,2,\Lambda,n)$ is the target model of normalized pixels, the center point is 0, and the probability density $q_u(x)$ of the target model can be expressed as:

$$q_u(x)=C\sum_{i-1}^{n}k\left(\left\|x_i^*\right\|^2\right)\delta[b(x_i)-u]. \tag{1}$$

In which δ is dirac function, $b(x_i)$ is index function, which reflects bin sequence of pixel x_i, C is normalized correlation coefficient, $k(*)$ is kernel function, and the mathematical expression of kernel function in this paper is shown as follows:

$$k(*)=k\left(\left\|\frac{x_0-x_i}{h}\right\|^2\right). \tag{2}$$

Too small or too large kernel function bandwidth will have different degrees of errors in the state description of the system, filtering out most of the target state information. In this paper, the scale estimation algorithm uses ellipse feature to adjust the parameters of kernel function $k(*)$, so as to adjust the target scale and approach the real state of the target to the greatest extent. Because this paper aims to monitor and track the target pedestrians, the traditional Mean Shift uses circles to search the target in the region, and the appearance model of pedestrians is closer to ellipse, so the ellipse feature method is used to estimate the size of the target model. The convergence diagram is shown in Figure 6:

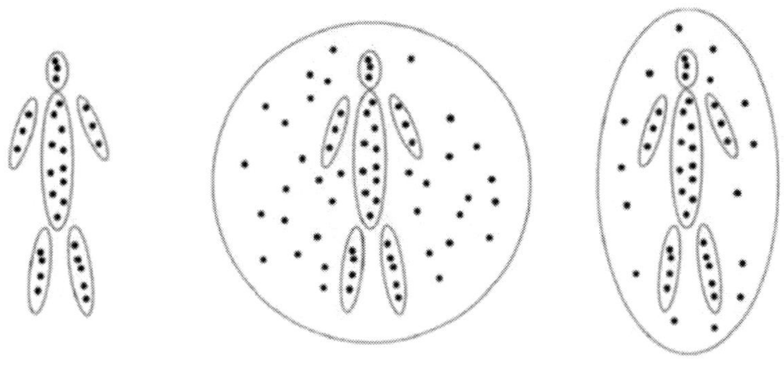

Figure 6. Convergence diagram

Let $\{x_i\}(i=1,2,\Lambda\ ,n)$ be the target sample point, and define the change of target scale as the same direction. $y=(y^1,y^2)$, $x_i=(x_i^1,x_i^2)$ the location of pixels in the target area will be estimated, and the probability density of the target model is:

$$q_u(x)=C\sum_{i=1}^{n}k\left(\frac{x_i^1}{a^2}+\frac{x_i^2}{b}\right)\delta[b(x_i)-u]. \tag{3}$$

The values of a and b are set to be half of the rectangular side length of the target area, and the parameter equation of ellipse is introduced into this function to modify the parameters of the function. The selected kernel function bandwidth is h, that is, the scale value of the current state of the target, and y_0 is set as the central area of the target. The probability density $p_u(x)$ of the candidate target model is as follows:

$$p_u(y_0,h)=C_h\sum_{i=1}^{n}k\left(\frac{(y_0^1-x_i^1)^2}{a^2h^2}+\frac{(y_0^2-x_i^2)^2}{b^2h^2}\right)\delta[b(x_i)-u]. \tag{4}$$

Since C_h is a normalized function, its parameter equation is as follows:

$$C_h=\frac{1}{\sum_{i=1}^{n}k\left(\frac{(y^1-x_i^1)^2}{a^2b^2}+\frac{(y^2-x_i^2)^2}{a^2b^2}\right)}. \tag{5}$$

5.2. Target scale estimation

The similarity between the target template and the template to be tested is estimated, and the target position and scale of the previous frame are estimated as the starting state of the target of the next frame until the target is collected, so as to achieve the purpose of self-adaptive adjustment of the scale. imilar to the traditional Mean Shift algorithm, the similarity between them is measured by Babbitt coefficient, as shown below:

$$\rho[P_u(y,h),q]\approx\rho(y,h)=\sum_{i=1}^{n}\sqrt{C_h\frac{h_0^2}{h^2}\sum_{i=1}^{n}k\left(\frac{(y_0^1-x_i^1)^2}{a^2h^2}+\frac{(y_0^2-x_i^2)^2}{b^2h^2}\right)\delta[b(x_i)-u]q_u}. \tag{6}$$

w_i is defined as:

$$w_i = \sum_{i=1}^{n} \sqrt{q_u / p_u(y_0)} \delta[b(x_i) - u] \tag{7}$$

The traditional Mean Shift method for solving the target position is to calculate the moving vector l of the target sample from the initial position to the actual position, as follows:

$$L(y_0, h_0) = \frac{\sum_{i=1}^{n} x_i w_i g\left(\frac{(y_0^1 - x_i^1)^2}{a^2 h^2} + \frac{(y_0^2 - x_i^2)^2}{b^2 h^2}\right)}{Y} - y_0 . \tag{8}$$

Among them, the parameter equation of y is:

$$Y = \sum_{i=1}^{n} w_i g\left(\frac{(y_0^1 - x_i^1)^2}{a^2 h^2} + \frac{(y_0^2 - x_i^2)^2}{b^2 h^2}\right). \tag{9}$$

It can be seen from the above that the current estimated position A and the estimated value B of the scale of the target are:

$$\left.\begin{aligned} y_1^1 &= \frac{1}{a^2} L^1(y_0, h) + y_0^1 \\ y_1^2 &= \frac{1}{b^2} L^2(y_0, h) + y_0^2 \end{aligned}\right\} . \tag{10}$$

$$h_1 = \left[1 - \frac{\sum_{i=1}^{n} w_i k\left(\frac{(y_0^1 - x_i^1)^2}{a^2 h^2} + \frac{(y_0^2 - x_i^2)^2}{b^2 h^2}\right)}{Y} \right] h_0 + \frac{\sum_{i=1}^{n} w_i \left(\frac{(y_0^1 - x_i^1)^2}{a^2} + \frac{(y_0^2 - x_i^2)^2}{b^2}\right) g\left(\frac{(y_0^1 - x_i^1)^2}{a^2 h^2} + \frac{(y_0^2 - x_i^2)^2}{b^2 h^2}\right)}{h_0 Y} \tag{11}$$

5.3. Design of Mean Shift Algorithm for Scale Estimation

In order to verify the effectiveness of the multi-sensor data fusion algorithm of patrol robot, a large number of safety early warning tests were carried out in the laboratory by using patrol robot by adjusting and controlling different temperature, humidity and gas leakage in the experimental warehouse. In order to facilitate verification, the safety degree of 30 groups of data in the test process was judged manually according to safety rules, and the robot fusion output value y ($0 \le y \le 1$) was taken as the robot early warning basis. It was stipulated that the closer y value was to 0, the higher safety degree was, on the contrary, the closer it was to 1, the lower safety degree was. And set no alarm when $y \le 0.3$, yellow alarm when $0.3 < y \le 0.7$ and red alarm when $y > 0.7$, with 30 sets of test data listed in Table 1.

Table 1. 30 sets of test data sheets.

serial number	relative humidity/ %	Air temperature/℃	CO concentration/ppm	Expected value of alarm	Expected judgment result	Robot fusion result value	Robot decision node
1	74	32	30	0	safe	0.004	safe
2	50	25	23	0	safe	0.0029	safe
3	77	50	14	0	safe	0.003	safe

4	42	10	15	0	safe	0.0193	safe
5	46	22	35	0	safe	0.1089	safe
6	80	19	20	0	safe	0.0001	safe
7	45	33	22	0	safe	0.0248	safe
8	56	32	12	0	safe	0.0001	safe
9	95	42	1	0	safe	0.0002	safe
10	44	22	3	0	safe	0.0004	safe
11	72	22	2	0	safe	0	safe
12	57	0	20	0	safe	0.002	safe
13	44	41	0	0	safe	0.0168	safe
14	40	88	22	1	danger	1	red alert
15	52	28	120	1	danger	0.9999	red alert
16	30	6	15	1	danger	0.9997	red alert
17	28	11	0	1	danger	0.9849	red alert
18	55	120	2	1	danger	0.9999	red alert
19	42	78	11	1	danger	1	red alert
20	33	55	25	1	danger	0.9994	red alert
21	50	60	11	1	danger	0.974	red alert
22	50	44	55	1	danger	0.9995	red alert
23	40	66	23	1	danger	0.9999	red alert
24	27	22	34	1	danger	0.9999	red alert
25	55	28	52	1	danger	0.981	red alert
26	22	33	1	1	danger	0.998	red alert
27	46	21	41	1	danger	0.9873	red alert
28	55	90	36	1	danger	0.9999	red alert
29	37	150	12	1	danger	1	red alert
30	44	31	98	1	danger	1	red alert

It can be seen from Table 2 that although the fusion results of 30 sets of test data by the robot have certain deviation from the alarm expectation, they are all within the allowable error range. The automatic judgment results generated according to the rules are consistent with the manual judgment results, which can be fully applied to the hierarchical early warning of the robot patrol process.

5.4. Analysis of Scale Estimation Algorithm

In this experiment, according to the different moving modes of the target in the camera range, two groups of experiments are taken to verify it. When the target moves in the shooting range, there are two moving situations. The farther away the target is from the camera, the smaller the target size is. On the contrary, the target size is getting bigger and bigger. The monitoring equipment used this time is 20 m, and the focal distance is fixed. The size of the selected sequence is 302×412 pixels, the length is 60 frames, and the frame rate is 30fps. When tracking the target, the tracking window of Mean Shift algorithm is fixed with the change of target scale. The scale change algorithm adopted in this paper improves the circular iteration of Mean Shift in searching for sample points. This paper adopts an approximate elliptical search method, which is closer to the appearance of human beings. With the change of target size, the tracking window is adjusted adaptively, which optimizes the disadvantages of traditional Mean Shift algorithm. That is to say, when the size of the target becomes larger, the local maxima affecting sample particles in some sample points will be ignored, which will lead to errors in tracking results; When the target size becomes smaller, some non-sample points are included in the search area, which increases the calculation amount, and will cause the non-sample points to have larger weights, resulting in the convergence error of sample points. Therefore, according to the distance between the target and the acquisition equipment, the size of the target changes obviously. The algorithm in this paper adjusts the adaptive window according to the change of the target size, which proves the effectiveness of this algorithm. Compared with the traditional algorithm, its stability is better. This paper makes statistics on the error pixels at the center of the target in the above two groups of experiments, as shown in Figure 7.

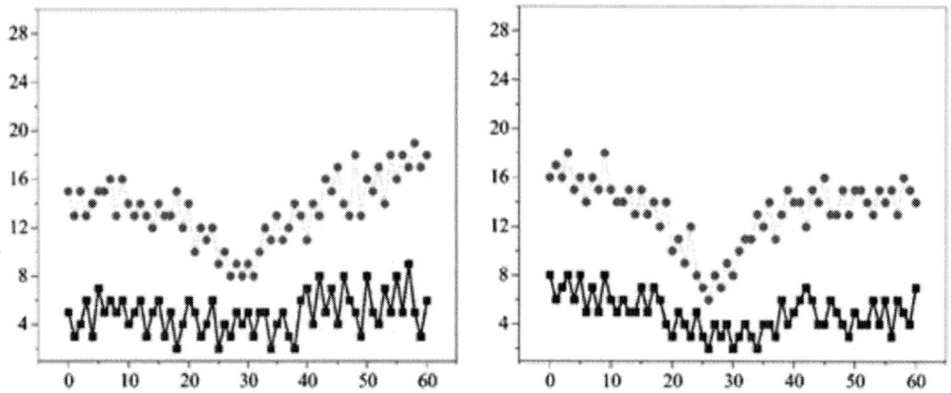

Figure 7. Target center point error pixel

It can be seen from the data graph that in the process of the target from far to near, at first, because the target size is smaller than the target template size, some target points are included in the convergence area, and weights are given for the next frame iteration, so the early error is larger, but as the target size gets closer and closer to the template size, the error gradually becomes smaller and enters normal tracking. With the moving target, the distance from the camera is getting closer and larger, and some sample points are excluded from the convergence area, which affects the local maximum of sample points. With the decreasing distance, more sample points are eliminated, and the convergence error increases. At about 30 frames, the tracking error is relatively small, mainly because the target template is closest to the fixed size in this time period, so the tracking effect is relatively ideal. The scale estimation algorithm used in this paper uses ellipse feature to adaptively adjust the convergence area of the target, which minimizes the error between the actual scale of the target and the tracking scale, and the tracking effect has been relatively stable, approaching the actual scale of the target to the greatest extent. In view of the above two moving modes of targets, this paper makes statistics on the target scale value in the moving process, and compares the actual target scale value in the Mean Shift fixed window with the scale value of the scale estimation algorithm adopted in this paper. The experimental results are shown in Figure 8.

IOP Conf. Series: Materials Science and Engineering 1179 (2021) 012002 doi:10.1088/1757-899X/1179/1/012002

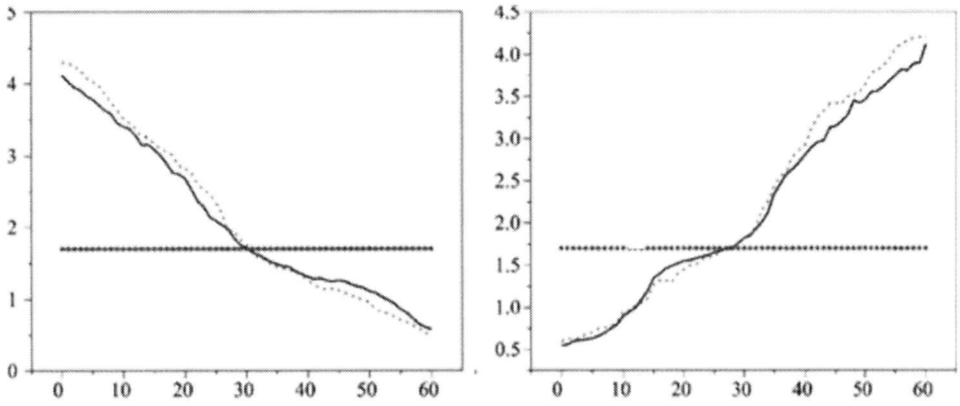

Figure 8. Hardware structure of environmental detection subsystem

6. Concluding remarks

Based on the application background of the safety monitoring of hazardous chemicals warehouse, this paper designs a robot which can patrol autonomously in the hazardous chemicals warehouse, and can use the environmental detection subsystem to carry out environmental detection data fusion and intelligent early warning during the patrol process. With the remote monitoring platform and cloud server, it realizes the remote monitoring and control function of managers to the hazardous chemicals warehouse. In this paper, scale estimation is used to improve the tracking drift phenomenon caused by the constant window of traditional Mean Shift, and the parameters of kernel function are adjusted by ellipse feature, so that the target search area and target template can be adjusted adaptively. In the tracking process, according to the change of the target's own scale, the tracking window will be adjusted adaptively. The algorithm is verified by two groups of experiments, and the effectiveness and stability of the method are proved by analyzing the pixel error in the center area of the target. The experimental results show that the autonomous patrol algorithm designed in this paper can traverse all the storage points of hazardous chemicals in a single island (only one middle obstacle zone) warehouse or no island, and the leakage points of hazardous chemicals can be detected quickly through the mounted environmental detection subsystem. At the same time, the mounted multi-sensor data fusion model can integrate the detection results of various sensors for data fusion processing, which can avoid false alarm and missing alarm in various situations, improve the accuracy and reliability of early warning of patrol robot system, and make the early warning mechanism of robot have certain intelligence.

Acknowledgments

Humanities and Social Sciences Research Project in Anhui Universities (SK2020A0527); Social Science Popularization Plan in Anhui Province (LZ201948); Anhui Quality Engineering Project (2019jyxm0468); Key Research Projects of Suzhou University (2019yzd17); Suzhou University School Quality Engineering Project (2018jyxm12).

References

[1] Cyrill S .Robotic mapping and exploration [M].Springer Tracts in Advanced Robotics, vol55, ISBN: 978-3-642-01096-5, 2019.
[2] Z. Fang, H. Liyan. A Survey of Multi-sensor Information Fusion Technology[J].Journal of Telemetry, Tracking and Command, 2016, 3: 000.
[3] D.A. Pomerleau Neural network perception for mobile robot guidance[M].Springer Science & Business Media, 2016.
[4] MA li,HA Park,CSG Lee.Closed-form inverse kinematic joint solution for humanoid robots[C]//Intelligent Robots and Systems (IROS), 2010 IEEE/RSJ International Conference on. IEEE, Taipei, Taiwan,010: 704-709.

[5] Di Giampaolo E Martinelli F.A passive UHF- RFID system for the localization of an indoor autonomous vehicle [J]. IEEE Transactions on Industrial Electronics, 2017, 59(10): 3961-3970.
[6] Xiao P, Luan YQ, Guo R, et al. Research of the laser navigation system for the intelligent patrol robot [J]. Automation & Instrumentation, 2018, 27(5): 5-9.
[7] Hess W, Kohler D, Rapp H, etal. Real-time loop closure in 2D LIDAR SLAM[C]//Robotics and Automation (ICRA),2016IEEE International Conference on. IEEE, 2016: 1271-1278.
[8] Qin H, Cong L, Sun X. Accuracy improvement of GPS/MEMS-INS integrated navigation system during GPS signal outage for land vehicle navigation [J]. Journal of Systems Engineering and Electronics, 2015, 23(2):256-264.
[9] J.B. Gao, C.J. Harris. Some remarks on Kalman filters for the multisensory fusion [J]. Information Fusion, 2017, 3:191-201.
[10]A. Weckenmann, X. Jiang. Multi sensor data fusion in dimensional metrology [J]. CIRP Annals-Manufacturing Technology, 2019, 58:701-721.
[11]Gerasimos G. Rigatos. Extended Kalman and Particle Filtering for sensor fusion in otion control of mobile robots [J]. Mathematics and Computers in Simulation, 2015, 81:590-607.
[12]Eva Besada-Portasa, Jose A. Lopez-Orozcoa, Juan Besada. Multi sensor fusion for linear control systems with asynchronous, Out-Of-Sequence and erroneous data [J]. Automatica, 2019,47:1399-1408.

Research on Multi-beam Handover Invitation Technology for GEO Mobile Satellite System

Yumeng Jiang

School of Electronic and Optical Engineering, Nanjing University of Science and Technology, Nanjing, Jiangsu, China

E-mail: yumeng@njust.edu.cn

Abstract. Different from the terrestrial cellular system, the GEO mobile satellite system has its own characteristics, such as the extremely large beams, the long propagation delay and the diversity of the MS (Mobile Station) velocity. Based on these characteristics, a cooperative beam-cluster is constructed, and a handover invitation technology based on the coordinated multi-beam is proposed in the paper. Using the sending mechanism of handover invitation by a variable sliding window, the technology can avoid the occurrence of the phenomenon of "ping pong" handover to realize seamless handover, and make the handover quick, efficient and reliable. Simulation results show that the algorithm significantly increases the handover success rate, prevents the unnecessary handoff and has a favourable adaptability to the high velocity MS.

1. Introduction

As the development of the next generation satellite mobile communication system towards higher throughput and transmission rate, the multi-beam system architecture is widely used, providing various services for terminals in the coverage area. When a communications user moves from one beam to another beam coverage area, beam handover management is required to ensure the continuity of communication [1]-[2].

Because it does not involve the issue of satellite selection, the current research on beam handover focuses on the allocation of channel resources and the handover guarantee strategy. For example: Literature [3] proposed that only when there are available channels in the current beam and the next switching beam, the call request can be accessed, and the call request can be queued according to the FIFO (First In First Out) principle; Literature [4] combines the handover guarantee algorithm with the LUI (Last Useful Instant) criterion, and queues according to the handover retention time to ensure that the most "critical" handover request is ranked first.

In recent years, a handover method based on inter-beam cooperation has also been proposed. Through negotiation between the current beam and candidate adjacent beams, the user terminal to be handed over is allocated random access channel resources in the adjacent beams. At this time, the terminal is maintaining the current beam, using the allocated channel resources to measure the link quality of the area, and finally determine the adjacent area that meets the handover requirements. Subsequently, the current beam initiates a handover request to the target beam for the user terminal that has obtained the synchronous information of adjacent beams. If the target beam accepts the request, the response information is returned to the current beam, and the current beam transmits the

IOP Conf. Series: Materials Science and Engineering 1179 (2021) 012003 doi:10.1088/1757-899X/1179/1/012003

information to the user through the handover command. Overall, the multi-beam cooperative handover technology has improved stability and reliability.

This paper introduces a handover invitation strategy based on multi-beam cooperation, which can increase the handover success rate.

2. Principle of multi-beam handover invitation Technology

Compared with the traditional handover method, which has limitations such as high handover failure rate, signaling loss and ping-pong handover, the invitation strategy allows the target beam to actively send handover invitations to mobile terminals with handover intentions in advance, changing the handover mechanism to passive be proactive, simplify the process of signal measurement and signal judgment, increase the switching speed, and reduce user service data interruption.

To implement the handover invitation strategy, it is necessary to first solve how the target beam sends a handover invitation to the communication user, including the sending time, sending method, and sending content. In addition, it is also important for communication user to solve the problem of how to deal with and answer the invitation.

2.1. Handover invitation sending mechanism based on sliding window

In order to achieve handover and automatic adaptation of the moving range, the window of sending the invitation is variable [5]. The window width W represents the number of beams in the window. As shown in figure 1.

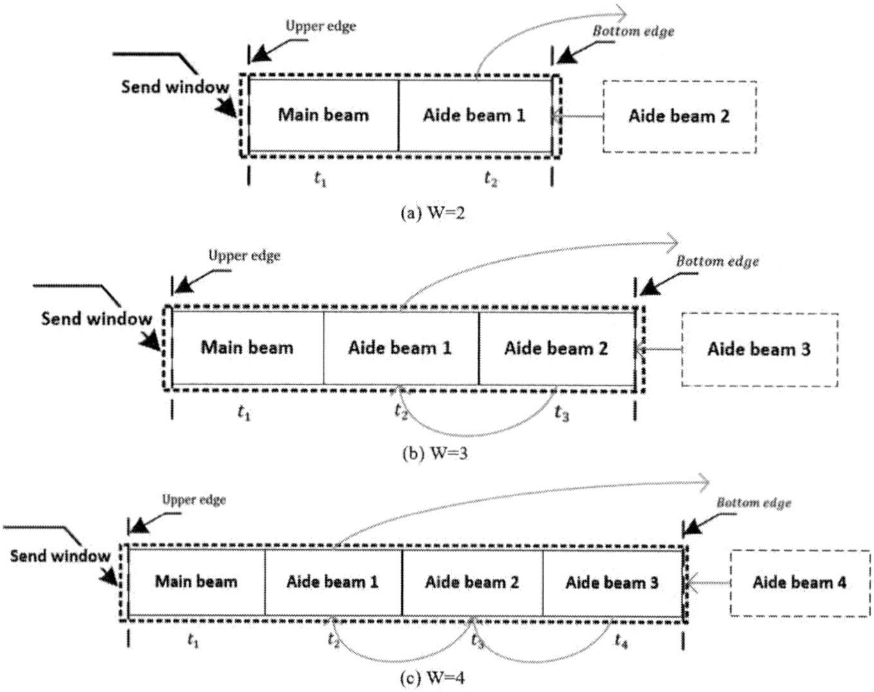

Figure 1. Handover invitation based on sliding window.

Each time a handover invitation is sent, the service beam (the main-beam) where the user is located, must always remain within the sending window and be at the upper edge of the window to ensure that the main beam signal reaches the user terminal first, effectively reducing the occurrence of "ping-pong handover". According to the sliding mechanism of the window, the order in which the aide-beams send invitations is determined by the signal quality that they reach the user.

Take W=2 as an example. When sending for the first time, the sending order is the main-beam and aide-beam 1 with the best signal quality; for the second sending, the main beam at the upper edge of the window remains unchanged, and aide-beam 2 with the second best quality is taken at the lower edge of the window. That is, the aide beam slides forward by one position, and the previously ranked aide-beam 1 moves to the end of the cooperative cluster. According to whether the user can correctly receive the handover invitation, it is determined whether the third or fourth transmission is required. The transmission order of the aide-beams is still arranged according to the previous sliding principle, until each beam in the cluster is traversed. Generally, when W=3, the actual effect is the best.

2.2. Multi-beam cooperation sends handover invitation

Determining a reasonable sending area can effectively improve the effectiveness of sending handover invitations [6]-[7]. Therefore, in this paper, according to the communication user's moving direction, the service beam and the position information of adjacent beams, the current beam and its adjacent beams are formed into a cooperative cluster [8]-[12]. Establish a handover invitation mechanism based on multi-beam cooperation.

The flow chart of multi-beam cooperation sending handover invitation to the user terminal is shown in Figure 2.

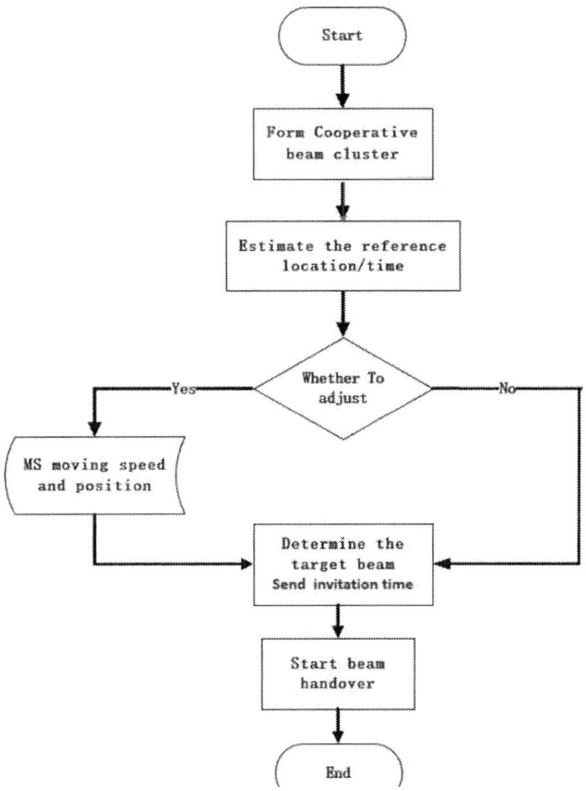

Figure 2. The flow chart of multi-beam cooperation sending handover invitation.

Step 1: Establish a cooperative cell, and the current beam of the communication user and the two adjacent beams form a cooperative relationship;

Step 2: According to the current movement trajectory of the communication user and estimate the reference position of the target beam to send the handover invitation;

Step 3: Combine the reference position for sending the invitation and half of the maximum overlap coverage distance of adjacent beam signals to predict the exact time for sending the handover invitation;

IOP Conf. Series: Materials Science and Engineering 1179 (2021) 012003 doi:10.1088/1757-899X/1179/1/012003

Step 4: According to the user's speed and the center position of the target beam, decide whether to adjust the time for sending the invitation, and finally determine the adjusted time as the time for sending the invitation;

Step 5: Start beam handover.

In the above process, the crucial issue is to estimate the reference location and reference time. The ideal time for the target beam to send the handover invitation is called the reference time, and the position of the MS at this time is the reference location. It is generally believed that the position where the coverage signal strength of the target beam and the current beam is equal is the reference location. The reference time can be obtained by the reference position and the moving speed of the MS.

Therefore, the communication environment model based on the handover invitation strategy of multi-beam cooperation is first established [13]-[14], as shown in Figure 3.

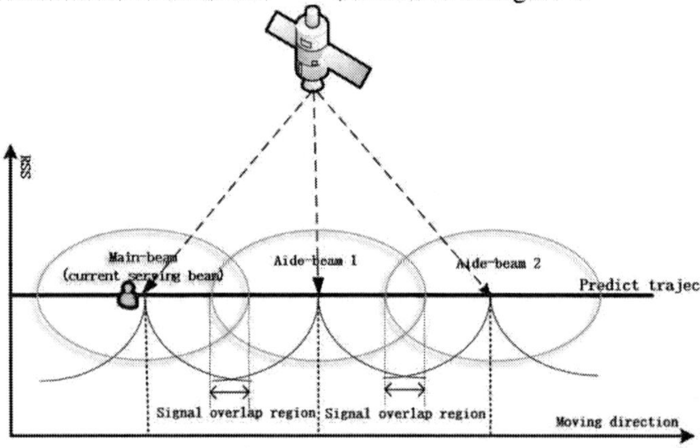

Figure 3. The communication environment model.

Estimate the RSS (Received Signal Strength) of MS (Mobile Station) according to the COST231-Hata loss prediction model.

When the MS moves away from the center of the Main-beam, the RSS gradually decreases in the form of a log-normal distribution. Assuming that distance between the center of main-beam and aide-beam 1 are D kilometres, the predicted moving direction of the MS is to move from the main-beam to the aide-beam 1. During the movement, the MS is d kilometres away from the center of the main-beam (current serving beam). At this time, the RSS from the current beam is denoted as $R_0(d)$, the RSS from aide-beam 1 and aide-beam 2 are $R_1(d)$ and $R_2(d)$. The received signal is averaged through an exponential filter window to eliminate Rayleigh fading. The impulse response of the filter is $f(d)$ and the average window width d_0. Then the expression of the received signal is:

$$\bar{R}(d) = f(d) * R(d) = 1/d_0 \cdot \int_0^{\infty} \exp{(-x/d_0)} z \cdot R(d-x)dx \tag{1}$$

Among them, the shadow fading components in the received signal are mutually independent zero-mean Gaussian random components. The variance is σ^2. $\bar{R}_0(d)$ and $\bar{R}_1(d)$ obey normal distribution. When $\bar{R}_0(d) = \bar{R}_1(d)$, the RSS by the MS from the main-beam and the aide-beam 1 is the same. It is considered that the MS is at the reference position of the aide-beam 1 (target beam) for sending the handover invitation (X_h, Y_h).

Use GPS to determine the current position (X, Y) of MS. The Doppler Shift is estimated according to the statistics of LCR (level crossing rates) or ZCR (zero crossing rates), thereby obtaining the approximate moving speed V of the MS.

Therefore, the reference time for sending the handover invitation can be expressed as:

$$T = \sqrt{(X - X_h)^2 + (Y - Y_h)^2}/V \tag{2}$$

In order to improve the success rate of beam handover, the time for sending the handover invitation can be advanced for a period of time. The target beam starts to send the handover invitation once MS passes through the signal overlap area. Set the lead time $T_0 = L/V$, Where L is half the distance of the beam signal overlap area, and the final time for the target beam to send the handover invitation is:

$$t = T - T_0 \tag{3}$$

3. Results and discussion

Figure 4 is a comparison of the simulation curves of the handover success rate of the two handover algorithms at different operating speeds. It can be seen from the figure that the handover success rate of the handover invitation algorithm is higher than that of the traditional handover algorithm. The handover success rate of the traditional handover algorithm has a significant downward trend, but the invitation algorithm declines more slowly.

Assuming that the distance between the center of the two beams D=3500km, the shadow fading variance $\sigma = 5dB$, the sampling interval $T_0 = 0.5s$, and the link fading threshold $\Delta = -95dB$.

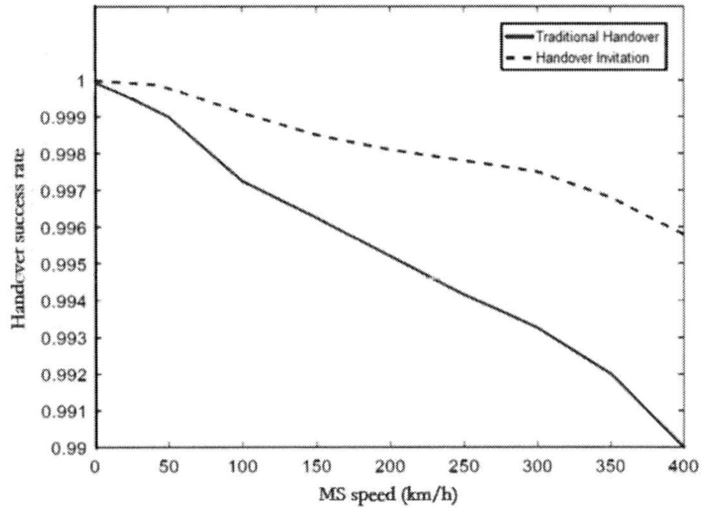

Figure 4. Comparison of simulation curves of handover success rate at different operating speeds.

4. Conclusion

Based on the current common beam handover technology research, this paper proposes a handover invitation mechanism based on inter-beam cooperation, and introduces the invitation sending mechanism based on sliding window and Multi-beam cooperation sends handover invitation in detail. MATLAB Simulation results prove the effectiveness of the technology. However, this technology needs to be established on the premise that the trajectory of MS is predictable. An interesting future step will be to solve how the target beam chooses an accurate invitation sending time when the trajectory is difficult to predict.

Acknowledgments

This work was supported by the National Natural Science Foundation of China under Grant 61771128.

References

[1] Song Li, Liu Aijun, Tian Xiaozhong, Xi Guo-zhong. Dynamic Handoff Algorithm for GEO Mobile Satellite System. Journal of System Simulation, 2009,21(11):3411-3415.

[2] He Yizhou. Research On New Genergation Mobile Satellite Communication System [D]. Beijing University of Posts and Telecommunications.

[3] E. Papapetrou, S. Karapantazis, G. Dimitriadis. Satellite Handover Techniques for LEO Networks [J]. International Journal of Satellite Communications and Networking, 2004, 3.

[4] E. Del Re, R. Fantacci, G. Giambene. Handover Queuing Strategies with Dynamic and Fixed Channel Allocation Techniques in Low Earth Orbit Mobile Satellite Systems [J]. IEEE Transactions on Communications, 1999, 1(47):89-102.
[5] Antonopoulos Chris G, Shang Yilun. Opinion formation in multiplex networks with general initial distributions [J]. Scientific Reports Volume 8, Issue 1. 2018. PP 2852.
[6] Wen Qiang, Li Jiying, Yang Yonghong, Li YA. The Research of LTE-R Handoff Algorithm Based on Fuzzy Prediction [C]. 2019 The 11th International Conference on Measuring Technology and Mechatronics Automation, 2019.
[7] Kang Guihua, Kang Hongbo, Pan Kunbei, Zhang Feng. The trigger mechanism of the quasiperiodic hand-over based on the handover invitation [P] .China patent, June 10, 2015-6-10.
[8] Patrick Marsch, Gerhard P Fettweis. Coordinated MultiPoint in Mobile Communications from Theory to Practice [M]. UK, Cambridge University Press, 2011.
[9] Re E D, Fantacci R, Giambene G. Different queuing policies for handover requests in low Earth orbit mobile satellite systems [J]. IEEE Transactions on Vehicular Technology, 1999, 48(2): 448-458.
[10] Qing Guo, Zhen yongwang, Xuemai Gu4t Satellite Communication System and Technique, University of Posts and Telecommunication Press, CHN, 2003.
[11] H.W. Kim, T.C. Hong, K. Kang, B.J. Ku, S. Kim and S.Yeo4t.A Satellite Radio Interface for IMT Advanced System Using OFDM, Information and Communication Technology Convergence, 109-124, 2010.
[12] Fei Yang, Meiyu Huang, Sihai Zhang and Wuyang Zhou, Multi-Layer Beam Architecture in Mobile Satellite Communications, 5, 6th International ICST Conference on Communications and Networking in China, CHINACOM 2011, pp: 1115.
[13] Chen Peng, Mi Gensuo, Luo Miao. Research on LTE-R Improved Switching Algorithm in High-speed Railway [J]. Railway Standard Design, 2018, 62(5): 150-154.
[14] Hyun—Ho Choi, Jong Bu Lim, Hyosun Hwang. Optimal Handover Decision Algorithm for Throughput Enhancement in Cooperative Cellular Networks [C] . IEEE 72nd Vehicular Technology Conference 2010 (1): 1-5.

AUTHOR INDEX